Life Science

____ Grade 3 ____

Written by Tracy Bellaire

The experiments in this book fall under eleven topics that relate to two aspects of life science: **Growth & Changes in Plants; and Animal Life Cycles**. In each section you will find teacher notes designed to provide you guidance with the learning intention, the success criteria, materials needed, a lesson outline, as well as provide some insight on what results to expect when the experiments are conducted. Suggestions for differentiation are also included so that all students can be successful in the learning environment.

Tracy Bellaire is an experienced teacher who continues to be involved in various levels of education in her role as Differentiated Learning Resource Teacher in an elementary school in Ontario. She enjoys creating educational materials for all types of learners, and providing tools for teachers to further develop their skill set in the classroom. She hopes that these lessons help all to discover their love of science!

Published in Canada by:
On The Mark Press
Belleville, ON
www.onthemarkpress.com

Funded by the
Government
of Canada

OTM2162   ISBN: 9781487710248
© On The Mark Press

At A Glance

Learning Intentions

Learning Intentions	Plants	Plant Parts	What Do Plants Need?	How Plants Grow	Importance of Plants	Aboriginal People and Plants	Growing Plants for Food	Invertebrates	Birds	Fish, Reptiles, Amphibians	Mammals
Knowledge and Understanding Content											
Recognize what a plant is and describe their physical characteristics	•	•									
Identify the main parts of a plant and explain their functions in helping a plant survive		•									
Conduct experimental inquiries to determine the basic needs of plants to grow healthy			•								
Investigate and describe seed germination, plant maturation, and special adaptations that help them survive				•							
Recognize the role plants play as a source of food energy for living things; construct a plant growing environment					•						
Describe the importance of some plants and their usage by First Nation's people						•					
Identify the ways plants are grown for food, recognizing the advantages and disadvantages of organically grown food, and describing harvesting methods of plants							•				
Research and describe the physical characteristics and behaviors of invertebrates								•			
Research and describe the physical characteristics and behaviors of birds									•		
Describe the physical characteristics of fish, reptiles, and amphibians; research to learn more about their life cycles										•	
Identify different mammals and describe their physical characteristics; research to learn more about a mammal											•
Thinking Skills and Investigation Process											
Make predictions, formulate questions, and plan an investigation		•	•	•	•						
Gather and record observations and findings using drawings, tables, written descriptions	•	•	•	•	•	•	•	•	•	•	•
Recognize and apply safety procedures in the classroom	•	•	•	•	•	•	•	•	•	•	•
Communication											
Assess different ways in which plants are important in the lives of people and other living things					•	•					
Application of Knowledge and Skills to Society and the Environment											
Assess different ways in which plants are important in the lives of people and other living things					•	•					
Assess the harmful effects plants face and determine how to minimize these effects, including a personal action plan					•						

OTM2162 ISBN: 9781487710248
© On The Mark Press

TABLE OF CONTENTS

AT A GLANCE — 2

TABLE OF CONTENTS — 3

TEACHER ASSESSMENT RUBRIC — 4

STUDENT SELF-ASSESSMENT RUBRIC — 5

INTRODUCTION — 6

PLANTS
Teacher Notes — 7
Student Activities — 9

PLANT PARTS
Teacher Notes — 15
Student Activities — 17

WHAT DO PLANTS NEED?
Teacher Notes — 30
Student Activities — 32

HOW PLANTS GROW
Teacher Notes — 45
Student Activities — 47

IMPORTANCE OF PLANTS
Teacher Notes — 57
Student Activities — 59

ABORIGINAL PEOPLE AND PLANTS
Teacher Notes — 67
Student Activities — 68

GROWING PLANTS FOR FOOD
Teacher Notes — 70
Student Activities — 71

INVERTEBRATES
Teacher Notes — 75
Student Activities — 76

BIRDS
Teacher Notes — 81
Student Activities — 82

FISH, REPTILES, AMPHIBIANS
Teacher Notes — 85
Student Activities — 86

MAMMALS
Teacher Notes — 92
Student Activities — 93

OTM2162   ISBN: 9781487710248
© On The Mark Press

Teacher Assessment Rubric

Student's Name: _____   Date: _____

Success Criteria	Level 1	Level 2	Level 3	Level 4
Knowledge and Understanding Content				
Demonstrate an understanding of the concepts, ideas, terminology definitions, procedures and the safe use of equipment and materials	Demonstrates limited knowledge and understanding of the content	Demonstrates some knowledge and understanding of the content	Demonstrates considerable knowledge and understanding of the content	Demonstrates thorough knowledge and understanding of the content
Thinking Skills and Investigation Process				
Develop hypothesis, formulate questions, select strategies, plan an investigation	Uses planning and critical thinking skills with limited effectiveness	Uses planning and critical thinking skills with some effectiveness	Uses planning and critical thinking skills with considerable effectiveness	Uses planning and critical thinking skills with a high degree of effectiveness
Gather and record data, and make observations, using safety equipment	Uses investigative processing skills with limited effectiveness	Uses investigative processing skills with some effectiveness	Uses investigative processing skills with considerable effectiveness	Uses investigative processing skills with a high degree of effectiveness
Communication				
Organize and communicate ideas and information in oral, visual, and/or written forms	Organizes and communicates ideas and information with limited effectiveness	Organizes and communicates ideas and information with some effectiveness	Organizes and communicates ideas and information with considerable effectiveness	Organizes and communicates ideas and information with a high degree of effectiveness
Use science and technology vocabulary in the communication of ideas and information	Uses vocabulary and terminology with limited effectiveness	Uses vocabulary and terminology with some effectiveness	Uses vocabulary and terminology with considerable effectiveness	Uses vocabulary and terminology with a high degree of effectiveness
Application of Knowledge and Skills to Society and Environment				
Apply knowledge and skills to make connections between science and technology to society and the environment	Makes connections with limited effectiveness	Makes connections with some effectiveness	Makes connections with considerable effectiveness	Makes connections with a high degree of effectiveness
Propose action plans to address problems relating to science and technology, society, and environment	Proposes action plans with limited effectiveness	Proposes action plans with some effectiveness	Proposes action plans with considerable effectiveness	Proposes action plans with a high degree of effectiveness

OTM2162   ISBN: 9781487710248
© On The Mark Press

Student Self Assessment Rubric

Name: _____ Date: _____

Put a check mark ✔ in the box that best describes you:

	Always	Almost Always	Sometimes	Needs Improvement
I am a good listener.				
I followed the directions.				
I stayed on task and finished on time.				
I remembered safety.				
My writing is neat.				
My pictures are neat and colored.				
I reported the results of my experiment.				
I discussed the results of my experiment.				
I know what I am good at.				
I know what I need to work on.				

1. I liked _____

_____

2. I learned _____

_____

3. I want to learn more about _____

_____

INTRODUCTION

The activities in this book have two intentions: to teach concepts related to life science and to provide students the opportunity to apply necessary skills needed for mastery of science and technology curriculum objectives.

Throughout the experiments, the scientific method is used. The scientific method is an investigative process which follows five steps to guide students to discover if evidence supports a hypothesis.

1. **Consider a question to investigate.**
   For each experiment, a question is provided for students to consider. For example, "Can a plant live and grow without sunlight?"

2. **Predict what you think will happen.**
   A hypothesis is an educated guess about the answer to the question being investigated. For example, "I believe that a plant will grow faster in the sunlight". A group discussion is ideal at this point.

3. **Create a plan or procedure to investigate the hypothesis.**
   The plan will include a list of materials and a list of steps to follow. It forms the "experiment".

4. **Record all the observations of the investigation.**
   Results may be recorded in written, table, or picture form.

5. **Draw a conclusion.**
   Do the results support the hypothesis? Encourage students to share their conclusions with their classmates, or in a large group discussion format.

The experiments in this book fall under 11 topics that relate to these aspects of life science: **Growth and Change in Plants, Invertebrates and Mammals.** In each section you will find teacher notes designed to provide you guidance with the learning intention, the success criteria, materials needed, a lesson outline, as well as provide some insight on what results to expect when the experiments are conducted. Suggestions for differentiation are also included so that all students can be successful in the learning environment.

ASSESSMENT AND EVALUATION:

Students can complete the Student Self-Assessment Rubric in order to determine their own strengths and areas for improvement. Assessment can be determined by observation of student participation in the investigation process. The classroom teacher can refer to the Teacher Assessment Rubric and complete it for each student to determine if the success criteria outlined in the lesson plan has been achieved. Determining an overall level of success for evaluation purposes can be done by viewing each student's rubric to see what level of achievement predominantly appears throughout the rubric.

OTM2162   ISBN: 9781487710248
© On The Mark Press

PLANTS

LEARNING INTENTION:

Students will learn about what a plant is, and about their physical characteristics.

SUCCESS CRITERIA:

- recognize a variety of plants in your neighborhood
- compare a variety of plants according to their physical characteristics
- identify the main physical characteristics of a plant
- describe the function of each of the main parts of a plant
- research a plant to determine its name, type, and label its parts

MATERIALS NEEDED:

- a copy of "What is a Plant?" worksheets 1 and 2 for each student
- a copy of "Plants in My Neighborhood" worksheet 3 for each student
- a copy of "The Sum of All Parts" worksheets 4 and 5 for each student
- a copy of "Pick a Plant, Any Plant!" worksheet 6 for each student
- iPods or iPads (optional)
- 5 or 6 assorted plants, soil, and planters
- access to the internet or local library
- chart paper, markers, pencil crayons, clipboards, pencils

PROCEDURE:

*This lesson can be done as one long lesson, or can be done in four or five shorter lessons.

1. Using worksheets 1 and 2, do a shared reading activity with the students. This will allow for reading practice and breaking down word parts to read the larger words. Along with the content, discussion of vocabulary words would be of benefit for their comprehension.

   Some interesting vocabulary words to focus on are:

   - categorized
   - flexible
   - rely
   - erosion
   - biennial
   - coniferous
   - herbs
   - water conditions
   - shed
   - annual
   - deciduous
   - herbaceous
   - soil conditions
   - bogs
   - perwial

2. Give students worksheet 3, a clipboard and a pencil. Take them out into the neighborhood to look for different kinds of plants. Encourage students to notice the types of trees, shrubs, grasses, flowered plants, etc. that are growing locally. *An option is to give students iPods or iPads to take photos of the different vegetation that they see.*

3. Display 5 or 6 assorted plants (roots exposed). Engage students in a discussion about the physical characteristics of each plant type. How are they different? What do they all have in common? Ask students to look back at the plants that they drew on worksheet 1 (or have taken photos of). How are they different? How are they the same? (Some common characteristics that should be noted are that all plants have a root system, stem, leaves. They differ in size, shape, color, and some have flowers).

4. Divide students into pairs. Give them worksheet 4. They will engage in a "think-pair-share" activity to discuss and then record answers to the questions on the worksheet. A follow up option is to come back together as a large group to share responses. Student responses could be recorded on chart paper and displayed in the classroom for future reference.

5. Give students worksheet 5. Read through with the students about the main parts of a plant and each of their purposes. Along with the content, discussion of certain vocabulary words would be of benefit to ensure students' understanding of the concepts.

Some interesting vocabulary words to focus on are:

- chlorophyll
- attract
- nutrients
- absorb
- pollinate
- supports
- anchor
- minerals

6. Students will choose one plant that they drew on Worksheet 1 that they would like to know more about. Using Worksheet 6, and with access to the internet or local library, they will provide more detail about the plant that they chose.

DIFFERENTIATION:

Slower learners may benefit by working in a small group with teacher support to orally answer the questions on Worksheet 4. Group responses could be recorded by teacher on chart paper and later shared with the large group. An additional accommodation would be for these learners to be partnered with a strong peer to complete Worksheet 6. They could choose one plant from either of their first worksheet to research.

For enrichment, faster learners could plant and care for the 5 or 6 assorted plants (used in item #3). They could chart their growth by measuring the plants' heights, or amount of flower production each has. This could be done by creating a simple bar graph.

OTM2162   ISBN: 9781487710248
© On The Mark Press

What is a Plant?

There are many different kinds of plants in our world. Every plant can be categorized as either being a tree, a bush or shrub, a weed, a vine, moss, or a herbaceous plant. Plants grow on land or in the water.

Trees are usually the tallest of all plants. They grow on one thick stem, which is called a trunk, with many leaves and branches. They grow in almost all kinds of soil and water conditions. Trees are perennials. This means that their life cycle is longer than three years.

Did you know that a tree grows a new layer on the outside of its trunk every year? If you count a tree's rings, you can find out how old it is.

Some trees are deciduous and some trees are coniferous. Deciduous trees lose their leaves as the cold season approaches, and they grow new leaves when the warm weather returns. Coniferous trees keep their leaves or needles, and shed only the oldest leaves or needles throughout the year.

Oak, maple, and elm are examples of tree types that are deciduous.

Pine, cedar, and spruce are examples of tree types that are coniferous.

OTM2162   ISBN: 9781487710248
© On The Mark Press

Bushes and shrubs are also perennials. They are usually much shorter than trees and have many woody stems.

Some bushes and shrubs are deciduous and some are coniferous. They are usually planted to help stop soil erosion and are very popular in gardens and parks.

A rose bush is a deciduous bush that goes into a winter sleep called dormancy when the frost comes.

Mosses are tiny plants that grow on rocks, soil, on the bark of trees, in streams and in bogs. They have small stalks and leaves. They do not have true roots and rely on water for survival.

Vines are weak-stemmed, flexible plants that rely on other plants for support. They often wind around branches and other objects to hold themselves up. Vines can be deciduous or coniferous and may flower or bear fruit.

Herbaceous plants are mainly flowering plants. This group includes herbs, grasses, vegetable plants, and flowers that are usually planted in gardens.

Some herbaceous plants are annuals, which last only one growing season. Some are biennials, which last two growing seasons. Some are perennials, which last longer than three growing seasons.

OTM2162 ISBN: 9781487710248

Name:

Plants in My Neighborhood

Have you ever noticed what kinds of plants are in your neighborhood?

Take a walk around your neighborhood. In the box below, draw and label the plants that you see.

OTM2162   ISBN: 9781487710248
© On The Mark Press

Name:

The Sum of All Parts

Think **Pair** **Share**

With a partner, do some thinking and sharing of ideas about the questions below. Record your ideas.

"What does a plant use its roots for?"

_____
_____
_____
_____
_____
_____

"What does a plant use its stem for?"

_____
_____
_____
_____
_____
_____

"What does a plant use its leaves for?"

_____
_____
_____
_____
_____
_____

"What does a plant use its flowers for?"

_____
_____
_____
_____
_____
_____

OTM2162   ISBN: 9781487710248
© On The Mark Press

Name:

Fast Facts!

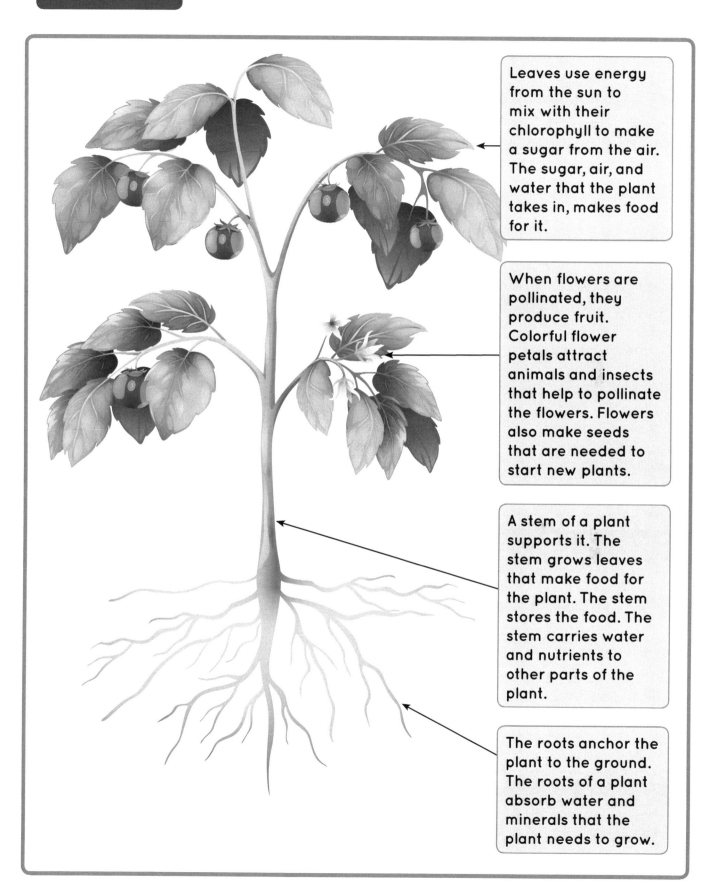

Leaves use energy from the sun to mix with their chlorophyll to make a sugar from the air. The sugar, air, and water that the plant takes in, makes food for it.

When flowers are pollinated, they produce fruit. Colorful flower petals attract animals and insects that help to pollinate the flowers. Flowers also make seeds that are needed to start new plants.

A stem of a plant supports it. The stem grows leaves that make food for the plant. The stem stores the food. The stem carries water and nutrients to other parts of the plant.

The roots anchor the plant to the ground. The roots of a plant absorb water and minerals that the plant needs to grow.

OTM2162   ISBN: 9781487710248

Pick a Plant, Any Plant!

Choose a plant that you saw in your neighborhood, and that you drew on Worksheet 1. Access the internet or visit your local library to help you answer these questions about this plant.

1. What is the name of this plant?

   _____

2. What type of plant is it? (circle one)

   tree   shrub/bush   moss   vine   herbaceous plant

3. Is this plant deciduous or coniferous?

   _____

4. Is this plant an annual, a biennial, or a perennial?

   _____

5. Draw your plant. Label its parts.

OTM2162   ISBN: 9781487710248

PLANT PARTS

LEARNING INTENTION:

Students will learn about how the main parts of a plant work to help it survive.

SUCCESS CRITERIA:

- describe the function of the root of a plant
- examine and identify plant roots as either tap or fibrous
- describe the function of the stem of a plant
- identify some edible plant stems
- describe the process of photosynthesis
- determine the different colors inside green leaves
- make observations and conclusions using written descriptions and illustrations
- identify the parts of a flower
- describe the ways the seeds of plants are distributed in nature
- identify some edible plant seeds
- label the parts of a tree plant and describe their purpose

MATERIALS NEEDED:

- a copy of "At the Root of It!" worksheet 1 and 2 for each student
- a copy of "Where It Stems From!" worksheet 3 and 4 for each student
- a copy of "Plant Leaves" worksheet 5 and 6 for each student
- a copy of "Discover the Color! worksheet 7, 8, and 9 for each student
- a copy of "The Flower" worksheet 10 for each student
- a copy of "The Seeds" worksheet 11 and 12 for each student
- a copy of "A Tree" worksheet 13 for each student

- about a dozen magnifying glasses
- four different plants with root systems exposed (a set for each group of students)
- 3 glass jars, a coffee filter, 3 labels, a shallow baking pan, a wooden spoon, a collection of three types of green leaves (for each group of students)
- water, a kettle, about 4 bottles of rubbing alcohol
- access to the internet, or local library
- scissors, chart paper, markers, pencils, pencil crayons
- sheets of white art paper, crayons, differently shaped leaves (*optional materials*)

PROCEDURE:

*This lesson can be done as one long lesson, or can be done in five or six shorter lessons.

1. Using worksheet 1, do a shared reading activity with the students. This will allow for reading practice and learning how to break down word parts in order to read the larger words in the text. Along with the content, discussion of certain vocabulary words would be of benefit for students to fully understand the passage.

   Some interesting vocabulary words to focus on are:

   - anchor
   - tap root
   - nutrients
   - strands
   - support
   - starch
   - storage
   - absorb
   - tropical
   - fibrous root

2. Divide students into small groups, and give each of them worksheet 2, four different plants with roots exposed, and magnifying glasses. Students will name and draw each plant, and determine if each plant has a tap or fibrous root system.

3. Give students worksheet 3. Read through as a large group to ensure students' understanding. Give students worksheet 4 to complete. They may need to visit a local library or access the internet to gather information about edible plant stems.

4. Using worksheets 5 and 6, do a shared reading activity with the students. This will allow for reading practice and learning how to break down word parts in order to read the larger words in the text. Along with the content, discussion of certain vocabulary words would be of benefit for students to fully understand the passage.

   Some interesting vocabulary words to focus on are:

   - trigger
   - broad
   - deciduous
   - evergreen
   - carbon dioxide
   - transpiration
   - chlorophyll
   - jagged
   - coniferous
   - function
   - oxygen
   - photosynthesis

5. Take students outside to collect 3 different types of green leaves (about a handful of each type). Upon returning to the classroom, divide students into small groups, and give each of them worksheets 7, 8, and 9, and the materials to conduct the investigation. *This investigation needs extended time to be completed, so starting early in the day is recommended. Students should conclude that the rubbing alcohol worked to break down the chlorophyll in the leaves so that their hidden colors could be revealed. This is relative to the reduction of warmth and sunlight in autumn, which results in a breakdown of the chlorophyll in nature's green leaves, to reveal their hidden colors before they die and fall from the deciduous trees.

6. Using worksheets 10 and 11, do a shared reading activity with the students. This will allow for reading practice and learning how to break down word parts in order to read the larger words in the text. Along with the content, discussion of certain vocabulary words would be of benefit for students to fully understand the passage.

Some interesting vocabulary words to focus on are:

- reproductive
- petals
- pistil
- scatter
- scent
- stamen
- nectar
- waterlogged
- producing
- anther
- fertilized
- excrement
- attract
- pollen
- distribute
- explode

7. Give students worksheet 12 to complete. They may need to visit a local library or access the internet to gather information about edible plant seeds.

8. Give students worksheet 13 to complete.

DIFFERENTIATION:

Slower learners may benefit by:

- working in a small group with teacher direction to complete worksheets 7, 8, and 9.

- working with a strong peer to conduct the research on edible stems and seeds on worksheets 4 and 12

- reducing the expectation to only finding two examples of edible plant stems and seeds on worksheets 4 and 12

For enrichment, faster learners could do a leaf rubbing art activity. Lay a leaf under a sheet of white art paper (on a flat surface). Using the side of a crayon, lightly rub the crayon over the paper so that a pattern of the leaf can be seen. Repeat with other leaves to create a design. Red, yellow, orange, brown, and green crayons could be used as a display of the colors that are seen in autumn. An additional step would be for these students to provide a short written explanation of why these colors appear in the leaves on some deciduous trees in autumn. Then, attach it to their art work.

16

OTM2162   ISBN: 9781487710248
© On The Mark Press

At the Root of It!

The roots of a plant anchor it to the ground. Roots are mostly underground and grow down into the soil. But, for some tropical trees, like the mangrove tree, some roots grow above the ground and help to support the tree.

Roots do not have leaves, but they often have root hairs which grow out into the soil to help absorb water and nutrients that a plant needs to grow. Did you know that some roots are storage areas and food for the plant? For example, the roots of beets, radishes, and carrots store food for the plant in the form of **starch**.

Roots are either tap or fibrous. **Tap roots** are roots that have one root that is longer than the rest. This larger root grows straight down. **Fibrous roots** have many strands of roots of similar size that spread out in all directions.

Beets, radishes, and carrots are examples of plants with tap roots.

Examples of plants with fibrous roots are lettuce and grass plants.

Examine the roots of four different plants. During your examination:

- name the plant

  draw the plant's root system

  determine its root system (circle tap or fibrous)

Name: _____

tap   fibrous

Name: _____

tap   fibrous

Name: _____

tap   fibrous

Name: _____

tap   fibrous

OTM2162   ISBN: 9781487710248
© On The Mark Press

Where It Stems From!

The stem of a plant is the part that usually has buds and leaves. Stems usually grow upwards and straight, but there are some plants like the strawberry plant, which have stems that grow along the ground. The upper part of a stem of a plant can bear branches, leaves, flowers, and fruit.

Stems have four main jobs:

1. They support the plant.

2. They grow leaves.

3. They carry water and nutrients from the roots to the leaves.

4. They provide food storage from some plants.

Did you know that some stems, like tree trunks, grow a new layer every year? The new layer becomes the outside of the stem. This new layer grows underneath the bark of the tree.

The new layer is the part of the stem that brings the food and water from the roots to the leaves.

As each year passes, the stem gets thicker and thicker as new layers are added.

OTM2162   ISBN: 9781487710248
© On The Mark Press

Did you know that some stems of plants are edible? If stems of plants provide food storage, water, and nutrients for itself, then some of those plants must be good for us to eat too!

Research It!

Find out the name of some edible plant stems. Draw and label them.

OTM2162 ISBN: 9781487710248
© On The Mark Press

Name:

Plant Leaves

There are many different kinds of leaves that come in many different shapes and sizes. Some leaves are long and thin, while others are broad and round.

Some leaves have downy hairs on their underside. Some leaves have jagged edges, while others have smooth edges. Some leaves even have a particular smell.

Leaves are often green in color because they have **chlorophyll** in them, but these leaves have hidden colors in them too!

The leaves on a Venus Fly Trap plant have stiff hairs called trigger hairs. Its trigger hairs sense insects that land on its leaves. The leaf snaps shut and traps the insect inside. It becomes food for the plant.

In autumn, the chlorophyll in these leaves breaks down and exposes the hidden colors in the leaves, such as yellow, orange, and red. The leaves then die and soon fall off. These leaves can be found on **deciduous** trees.

Coniferous trees, such as pine trees, have needle like leaves that remain green year round. Coniferous trees are also known as evergreens.

OTM2162   ISBN: 9781487710248
© On The Mark Press

Leaves take in sunlight. They have openings that let water and air in and out. Leaves breathe in **carbon dioxide** and breathe out **oxygen**. This is very important to humans. We need the oxygen that the plants provide. Leaves also pass water vapor into the air, this is called **transpiration**.

The leaves on a plant have an important function for the plant. The leaves produce food for it. This food making process is called **photosynthesis**. During photosynthesis, leaves get energy from the sun, and they use their chlorophyll to create a simple sugar from the air. The sugar and the water that the plant takes in make food for the plant.

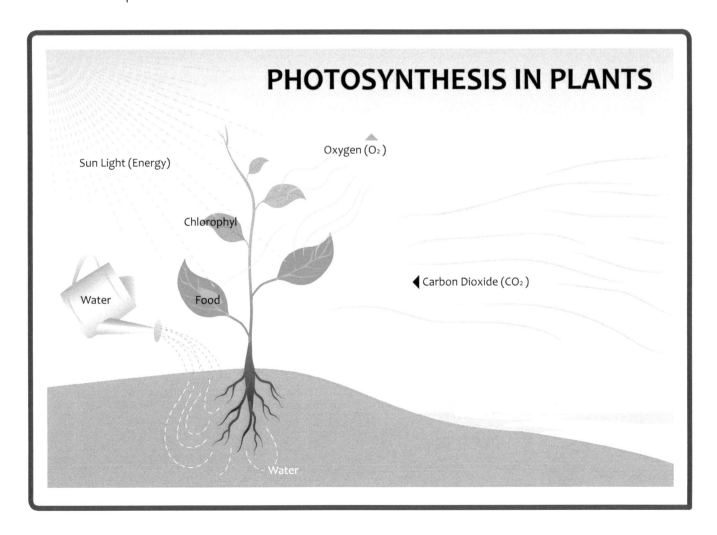

PHOTOSYNTHESIS IN PLANTS

Oxygen (O_2)

Sun Light (Energy)

Chlorophyl

Water

Food

Carbon Dioxide (CO_2)

Water

22

OTM2162 ISBN: 9781487710248

Discover the Color!

What hidden colors are in those leaves outside? Collect some green leaves and try this!

Materials Needed:

- 3 glass jars
- water
- a wooden spoon
- a pair of scissors
- a collection of three types of green leaves
- a shallow baking pan
- rubbing alcohol
- a kettle
- a marker
- a coffee filter
- 3 labels

What To Do:

1. Tear up each type of leaf into small pieces and place each type into its own jar. Label the jars with the type of leaf that is in it.
2. Carefully pour some rubbing alcohol into each of the jars, so that it just covers the leaves that are in it.
3. Using the wooden spoon, mix and grind up the leaves in the jars.
4. Place the glass jars into a shallow baking pan. **Your teacher will pour boiling water into the pan so that it covers the jars half way up.**
5. After 30 minutes, remove the jars from the hot water bath. Remove the leaves from the jars. Compare the shades of chlorophyll in the different leaf types. Record your observations on worksheet 8.
6. Cut the coffee filter into long strips. Place one strip into each glass jar so that it stands tall out of the water. Let it sit for 2 hours.
7. Pull out the strips. What do you notice? Record your observations on worksheets 8 and 9.
8. Make conclusions about the hidden colors in green leaves. Record them on worksheet 9.

Let's Observe

Part One

Describe the shades of chlorophyll in each of the jars.

Jar 1 - _____ leaves

_____

_____

_____

Jar 2 - _____ leaves

_____

_____

_____

Jar 3 - _____ leaves

_____

_____

_____

Part Two

What did you observe after you let the coffee filter paper sit in the rubbing alcohol for 2 hours?

_____

_____

_____

_____

_____

OTM2162 ISBN: 9781487710248
© On The Mark Press

Name:

Illustrate the colors that you saw.

Jar 1	Jar 2	Jar 3

Let's Conclude

What did the rubbing alcohol do to the leaves?

_____

_____

Explain what you have learned about the colors inside green leaves.

_____

_____

_____

OTM2162   ISBN: 9781487710248
© On The Mark Press

The Flower

Flowers are the prettiest parts of plants. They come in all shades and colors, and most of them smell very nice. Flowers also have an important job in helping a plant to survive. Let's learn more about this!

The flowers on most plants are the reproductive part. Flowers become pollinated and produce fruit, which creates the seeds to start a new plant. Some flowers go straight to seed without producing fruit, like the dandelion plant.

The petals on a plant are usually its largest part. The petals are colorful and attract animals and insects that help to pollinate the flower. The petals on a flower have a scent. The scent comes from a type of oil in the petals. The scent also attracts insects.

Inside the **petals** are the **stamens**. The stamens are long and thin. At the end of the stamen is the **anther** which produces the **pollen**.

In the very center of the flower is the **pistil**. This bottle shaped part of the flower contains **nectar** at the top and eggs at the bottom.

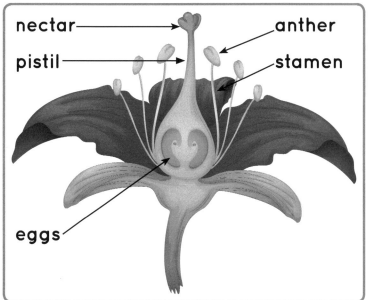

The **eggs** need to be fertilized by the pollen to produce **fruit** or **seeds**.

OTM2162 ISBN: 9781487710248
© On The Mark Press

The Seeds

Plants reproduce themselves using seeds. Seeds come in all shapes and sizes, from the smallest, which are like the head of a pin, to the largest, which can be the size of your hand.

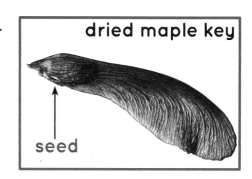

dried maple key

seed

Seeds contain an egg that needs to be fertilized to create a new plant. Once the egg is fertilized, the plant gets the seeds ready for travel. Plants distribute many seeds, sometimes several hundred. Yet most do not take root and grow a new plant. They may get broken, or get waterlogged, or rot. Some plants even get eaten, so the seeds get destroyed.

Plants distribute their seeds in different ways. Some seeds fly with the wind, like maple keys and dandelion seeds. Some seeds have hooks and bristles that attach themselves to animals and people who walk past them. Burrs are an example of seeds that travel this way.

Other seeds travel by floating on lakes, rivers, streams, or oceans. The coconut is a fruit that often falls into the ocean. It may travel to a new shore where it breaks open and the seed within takes root.

Some seeds are distributed by animals that eat the fruit of some plants. They pass the seeds as excrement.

There are plants, such as impatiens, that have seed pods which explode, causing their seeds to spray in all directions. Another example of an exploding seed pod is found on the rubber tree plant. Did you know that the fruit on the rubber tree will burst open when it is ripe, and scatter many seeds, up to about 30 meters from the tree?

This is a seed of a rubber tree plant.

OTM2162   ISBN: 9781487710248
© On The Mark Press

Name:

Research It!

Did you know that the seeds of some plants are edible? Access the internet or visit your local library to help you find out the name of some edible seeds. Draw and label them.

Have you eaten any of these seeds? Which ones?

_____

OTM2162   ISBN: 9781487710248
© On The Mark Press

Name:

A Tree

Use the words in the Word Box as labels to create a diagram of a tree.

roots

trunk

bark

leaves

seeds

Match the parts of a tree in column A to its purpose in column B.

Column A	Column B
Leaves	• bring water and nutrients from the ground
Trunk	• holds up the leaves, fruit, and seeds
Roots	• will grow to be new trees
Bark	• make food for the tree
Seeds	• protects the tree from disease and animals

OTM2162   ISBN: 9781487710248
© On The Mark Press

WHAT DO PLANTS NEED?

LEARNING INTENTION:

Students will learn about the basic needs of plants in order to grow healthy.

SUCCESS CRITERIA:

- make and record predictions about the needs of a plant
- conduct experiments to investigate the needs of a plant
- make and record observations using diagrams and written descriptions
- make conclusions and connections about the needs of plants using written descriptions

MATERIALS NEEDED:

- a copy of "Do Plants Need Light?" worksheet 1, 2, and 3 for each student
- a copy of "Heat Energy from the Sun" worksheet 4, 5, and 6 for each student
- a copy of "Do Water and Air Play a Part?" worksheet 7, 8, and 9 for each student
- a copy of "The Space Between" worksheet 10, 11, and 12 for each student
- a copy of "The Needs of a Plant" worksheet 13 for each student
- two small plants in planter containers, one labeled "light" and the other labeled "dark"
- two jars, two thermometers
- 4 small potted plants for each group of students
- medium sized boxes (one for each group of students)
- access to water, a few measuring cups, a couple of watering cans
- 8 small plants, 2 planter boxes, planting soil, a small garden shovel
- rulers, pencils, pencil crayons, markers, labels, sheets of poster paper

PROCEDURE:

*This lesson can be done as one long lesson, or can be done in four or five shorter lessons.

1. As a large group, conduct the experiment on worksheet 1. Give students worksheets 2 and 3. They will make a prediction about what may happen to a plant that is placed in direct sunlight and to a plant that is placed in a dark place. Over the next three weeks, students will make observations about the growth of both plants and record them. Then, they will make a conclusion based on their observations. (Sunlight is needed to make plants grow.)

2. As a large group, conduct the experiment on worksheet 4. Give students worksheets 5 and 6. They will make a prediction about the effect of direct sunlight on water. Using a thermometer, take the temperature of the water in each jar, and have students record this. Place one jar of water in direct sunlight and one away from sunlight. After two hours, have students feel the water in each jar, take the temperature of the water in each jar, and record it. They will make a conclusion and connections based on their observations. (Plants get warmth from the sun. Sunlight also warms up the soil and rainwater.)

3. Give students worksheets 7, 8, and 9. This experiment can be done as a large group, or in small groups. Students will make a prediction about what may happen to a plant that is placed in a dark place with no water or air, to one that is in a dark place with air and water, to one that is in sunlight but given no water, and to one that is in sunlight, in open air, and given water. Over the next two weeks, students will make observations and conclusions about the growth of the plants and record them on worksheets 8 and 9. Observation prompts: What color are the leaves? Does the plant look healthy or sick? (Sunlight, water, and air are needed to make plants grow healthy.)

OTM2162   ISBN: 9781487710248
© On The Mark Press

4. As a large group, conduct the experiment on worksheet 10. Give students worksheets 11 and 12. Over the next four weeks, students will make observations about the affect of adequate space on the growth of the plants. Then, they will make a conclusion based on their observations. (Space is needed by plants in order to grow healthy.)

5. Give students worksheet 13 to complete. (Sunlight, warmth, water, air, and space are needed for plants to grow healthy.)

*As an activity to enhance the learning about the physical characteristics and needs of plants, show students The Magic School Bus episode called "Gets Planted". Episodes can be accessed at www.youtube.com

DIFFERENTIATION:

Slower learners may benefit by working as a small group with teacher support to conduct the experiment on worksheets 7, 8, and 9 (if this experiment is not being completed as a large group activity). An additional accommodation would be for these to discuss their findings after each experiment within a small group with teacher support to consolidate their learning. This will allow more ease when having to complete worksheet 13.

For enrichment, faster learners could create a 'Gardening Tips' poster that defines the needs of a plant in order to grow healthy.

Do Plants Need Light?

Do plants really need sunlight to grow? Let's conduct an experiment to test if sunlight affects the growth of plants!

Question: Can a plant live and grow without sunlight?

Materials Needed:

• 2 small potted plants (one labeled "**LIGHT**" and the other labeled "**DARK**")

• water

• a measuring cup

• a ruler

• sunlight

What To Do:

1. Make a prediction about the answer to the question. Record it on worksheet 2.

2. Place the "**LIGHT**" plant in a sunny place, and place the "**DARK**" plant in a dark place.

3. Every 3 or 4 days, water the plants. Be sure to give them the same amount of water.

4. Every Monday, measure the height of each plant.

5. Using worksheet 2, record your observations of both plants every Monday, for 3 weeks.

6. Make a conclusion about what you observed.

OTM2162   ISBN: 9781487710248
© On The Mark Press

Let's Predict

Can a plant live and grow without sunlight?

_____

_____

Let's Investigate

This is what I saw each week.

WEEK 1

	Light Plant	Dark Plant
How tall is the plant?		
Does the plant look healthy or sick?		

WEEK 2

	Light Plant	Dark Plant
How tall is the plant?		
Does the plant look healthy or sick?		

WEEK 3

	Light Plant	Dark Plant
How tall is the plant?		
Does the plant look healthy or sick?		

OTM2162 ISBN: 9781487710248
© On The Mark Press

Draw and label a picture to show what happens when a plant is grown in:

sunlight	a dark place

Let's Conclude

Do plants need sunlight to live and grow? Explain.

_____

_____

_____

_____

OTM2162   ISBN: 9781487710248
© On The Mark Press

Heat Energy from the Sun

The sun is a source of heat energy. Let's conduct an experiment to test how sunlight affects water temperature.

Question: How does direct sunlight affect water temperature?

Materials Needed:

- 2 jars
- water
- a measuring cup
- 2 thermometers

What To Do:

1. Make a prediction about the answer to the question. Record it on worksheet 5.

2. Pour 200mL of water into both jars. Using a thermometer, take the temperature of the water in each jar. Record it on your worksheet.

3. Put one jar in direct sunlight and put one away from direct sunlight.

4. After 2 hours, feel the temperature of the water in each jar. Take the temperature of the water in each jar and record it on worksheet 5.

5. Make a conclusion about what you observed.

OTM2162 ISBN: 9781487710248
© On The Mark Press

35

Name:

Let's Predict

How does direct sunlight affect water temperature?

_____

_____

_____

Let's Investigate

In the beginning, the temperature of the water in each jar was:

Temperature of the water in Jar 1 (direct sunlight)	Temperature of the water in Jar 2 (not in sunlight)

After 2 hours, the temperature of the water in each jar was:

Temperature of the water in Jar 1 (direct sunlight)	Temperature of the water in Jar 2 (not in sunlight)

Let's Conclude

Was your prediction correct? Explain.

_____

_____

_____

36

OTM2162   ISBN: 9781487710248
© On The Mark Press

Let's Connect It!

Use the words in the Word Box to complete the sentences.

rise	warmer	warmer	heat

You learned that the temperature of the water in the jar that was in direct sunlight was _____ than the temperature of the water in the jar that was not in sunlight. The warmth from the sun caused the temperature to _____.

So an object, such as a plant that is in direct sunlight, would be _____ than a plant that is not in sunlight.

To connect this, you have learned that besides light, plants also get _____ from the sun.

Challenge Question:

Besides the plant itself that is in direct sunlight, what else is the sunshine warming that can have an effect on the growth of a plant?

_____

_____

_____

_____

_____

_____

OTM2162 ISBN: 9781487710248
© On The Mark Press

Do Water and Air Play a Part?

Question: Do water and air affect the growth of a plant?

Materials Needed:

- 4 small potted plants (one labeled "**DARK, AIR, and WATER**", one labeled "**DARK, NO AIR, NO WATER**", one labeled "**LIGHT and AIR**", and one labeled "**LIGHT, AIR, and WATER**")

- water

- watering can

- measuring cup

- a medium sized box

- a plastic bag

What To Do:

1. Make a prediction about the answer to the question. Record it on worksheet 8.

2. Place the "**DARK, AIR, and WATER**" plant in a place without light.

3. Place the "**DARK, NO AIR, NO WATER**" plant in a plastic bag and place it under the box.

4. Place the "**LIGHT and AIR**" and the "**LIGHT, AIR, and WATER**" plants in a sunny place.

5. Every 3 days, water the "**DARK, AIR, and WATER**" plant, and water the "**LIGHT, AIR, and WATER**" plant. Give them the same amount of water.

6. Record your observations of all plants at the end of each week, for 2 weeks.

7. Make conclusions about what you observed.

OTM2162 ISBN: 9781487710248
© On The Mark Press

Name:

Let's Predict

Do water and air affect the growth of a plant?

_____

_____

Let's Investigate

Describe what each plant looked like at the end of each week.

WEEK 1

DARK, AIR, and WATER	DARK, NO AIR, NO WATER	LIGHT and AIR	LIGHT, AIR, and WATER

WEEK 2

DARK, AIR, and WATER	DARK, NO AIR, NO WATER	LIGHT and AIR	LIGHT, AIR, and WATER

Drawing of the watered plant grown in the dark:	Drawing of the plant grown in the dark, with no air or water:
Drawing of the plant grown in sunlight, without water:	Drawing of the watered plant grown in open air and sunlight:

Let's Conclude

What do plants need to live and grow healthy and strong?

_____

_____

How do you know this?

_____

_____

OTM2162   ISBN: 9781487710248
© On The Mark Press

The Space Between

We all enjoy the freedom to move and the space to stretch out. Plants are not much different than us humans. Let's conduct an investigation to see just how much, plants like their own space!

Materials Needed:

- 8 small plants
- 2 planter boxes or large containers
- planting soil
- a small garden shovel

- water
- watering can
- measuring cup
- a ruler

What To Do:

1. Place some soil in the planter boxes.

2. Plant 4 of the plants in the first box, very close together.

3. Plant the other 4 plants in the second box with lots of space between them.

4. Place both planter boxes in a sunny place.

5. Every 3 days, water the plants. Be sure to give them the same amount of water.

6. Record your observations of the plants in both planter boxes at the end of each week, for 4 weeks.

7. Make conclusions about what you observed.

OTM2162   ISBN: 9781487710248
© On The Mark Press

Let's Investigate

Describe what the plants looked like at the end of each week.

WEEK 1

Plants with no space between them	Plants with space between them

WEEK 2

Plants with no space between them	Plants with space between them

OTM2162   ISBN: 9781487710248
© On The Mark Press

Describe what the plants looked like at the end of each week.

WEEK 3

Plants with no space between them	Plants with space between them

WEEK 4

Plants with no space between them	Plants with space between them

To conclude, do plants need space to grow and be healthy?

_____

_____

OTM2162   ISBN: 9781487710248
© On The Mark Press

Name:

The Needs of a Plant

Let's Review!

You have learned a lot about the needs of a plant. Show what you know by completing the web below with things plants need to grow healthy.

Things that plants need to grow healthy.

OTM2162   ISBN: 9781487710248
© On The Mark Press

HOW PLANTS GROW

LEARNING INTENTION:

Students will learn about seed germination, how plants grow to maturity, and special adaptations that help plants to survive.

SUCCESS CRITERIA:

- identify plants that grow from a seed and ones that grow from a bulb
- observe and illustrate the germination process of a bean seed
- track and graph the growth of a bean plant
- describe how plants germinate from a seed and grow into a mature plant
- compare the germination process of two different seeds
- identify and describe special adaptations of certain plants

MATERIALS NEEDED:

- a copy of "Germination" worksheet 1 for each student
- a copy of "Seeds or Bulbs?" worksheet 2 for each student
- a copy of "Grow Your Own Bean Plant!" worksheet 3, 4, 5 and 6 for each student
- a copy of "Germination Challenge!" worksheet 7 and 8 for each student
- a copy of "Plant Adaptations" worksheet 9 and 10 for each student
- access to the internet or local library
- about 5 bean seeds, 2 plastic Ziploc bags, 2 planters (for each student)
- planting soil, a few small garden shovels, access to water, a couple of watering cans, a few measuring tapes, a few spray bottles, paper towels

- other assorted seeds (e.g., pea, bean, zucchini, radish)
- rulers, pencils, pencil crayons, chart paper, markers
- seeds of different sizes and colors, glue, art paper *(optional materials)*

PROCEDURE:

*This lesson can be done as one long lesson, or be done in five or six shorter lessons.

1. Using worksheet 1, do a shared reading activity with the students. This will allow for reading practise and learning how to break down word parts in order to read the larger words in the text. Along with the content, discussion of certain vocabulary words would be of benefit for students to fully understand the passage.

   Some interesting vocabulary words to focus on are:

   - anchoring
   - minerals
   - ripened
   - pollinate
   - moisture
   - fruit
   - absorb
   - produce
   - swell
   - bearing
   - germination

2. Give students worksheet 2 to complete. They will need to visit a local library or access the internet to gather information about plants that grow from seeds and those that grow from bulbs.

3. Give students worksheets 3, 4, 5, and 6, and the materials to germinate their own bean seeds. Students will record observations of their beans' germination. Upon germination, the students will plant their bean seeds and monitor their plant's growth through graphing observations and forming conclusions. They will also create a diagram of its life cycle. *This investigation will span over a number of days.

OTM2162   ISBN: 9781487710248
© On The Mark Press

4. Give students worksheets 7 and 8, and the materials to germinate a seed type of their choice. Students will monitor and record observations of their seeds' germination process, then compare it to their bean plant's germination process. *This investigation will span over a number of days.

5. Using worksheet 9, do a shared reading activity with the students. This will allow for reading practise and learning how to break down word parts in order to read the larger words in the text. Along with the content, discussion of certain vocabulary words would be of benefit for students to fully understand the passage.

Some interesting vocabulary words to focus on are:

- photosynthesis
- climates
- energetic
- chemical
- environments
- adaptations
- predators
- distasteful
- dormancy
- deciduous
- poisonous
- allergic reaction

6. Give students worksheet 10 to complete. They will need to visit a local library or access the internet to gather information about certain plant adaptations. A follow up option is to have students orally share information about one plant's adaptation with a classmate or within a small group.

*As an activity to enhance the learning about how seeds germinate into mature plants, show students The Magic School Bus episode called "Goes to Seed". Episodes can be accessed at www.youtube.com

DIFFERENTIATION:

Slower learners may benefit by:

- working together as a small group with teacher support to germinate and track the growth of a bean plant using worksheets 3, 4, 5, and 6. This would produce one set of data to graph on the line graph. The creation of a life cycle diagram on worksheet 6 could be done together on a large sheet of chart paper. This would give opportunity for discussion about plant development.

- an elimination of the comparison of the seeds' germination process on worksheet 8

- a reduction of expectation by only researching for one plant adaptation on worksheet 10

For enrichment, faster learners could make a seed picture. Using seeds of assorted sizes and colors, students will plan out a design or picture on art paper (outdoor scene or geometric pattern). Then they can use glue to attach the seeds to the paper. Let the seed pictures dry overnight.

OTM2162 ISBN: 9781487710248
© On The Mark Press

Name:

Germination

Nearly all plants grow from seeds. These plants are pollinated either by another plant or by themselves. Seeds that are ripened fall to the ground where moisture loosens the seed cover. The seed absorbs moisture and begins to swell. It is at this point that germination can take place. **Germination** occurs when a seed begins to grow or spout.

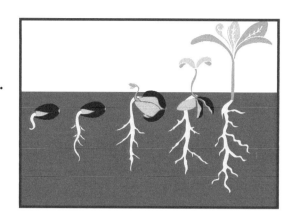

Once the seed begins germination, it sends a tiny stalk up and sends roots down into the soil. The roots draw water and minerals up into the seed to help it grow.

As the stalk grows, tiny leaves open up and begin making food for the plant. The leaves spread wide to absorb sunlight. The stalk grows taller, more leaves open and the roots continue to spread out, anchoring the plant to the ground.

If the plant is fruit-bearing, the young plant begins to grow buds in a couple of months. These buds open up and flowers appear. Insects and birds pollinate the flowers and the fruit begins to grow.

Most plants are grown from seeds, but some plants grow from bulbs. Bulbs usually grow underground and produce a stem and leaves that grow above ground. Common plants that are bulbs are onions, garlic, tulips, and daffodils.

OTM2162 ISBN: 9781487710248
© On The Mark Press

Name:

Seeds or Bulbs?

Research It!

Access the internet or visit your local library to help you find out the name of some plants that grow from seeds and some that grow from bulbs.

Grow from a seed	Grow from a bulb

Have you eaten any of these plants? Which ones?

_____

_____

_____

OTM2162   ISBN: 9781487710248
© On The Mark Press

Name:

Grow Your Own Bean Plant!

Have you ever wondered how a seed grows into a plant? In this experiment you will watch a bean seed grow into a plant. Let's grow!

Materials Needed:

- about 5 bean seeds
- a plastic Ziploc bag
- spray bottle
- a paper towel
- a planter

- water
- watering can
- a measuring tape
- a small gardening shovel
- soil

What To Do:

1. Using the spray bottle of water, mist the paper towel on both sides.

2. Put the bean seeds on the paper towel and fold it over a couple of times so that the bean seeds are covered.

3. Place the paper towel with the seeds in it, into the plastic Ziploc bag, and seal it. Leave it for 3 days.

4. After 3 days, check on your seeds. Has there been any change? On worksheet 3, record your observations.

5. Use the spray bottle to mist the paper towel, then place it back into the Ziploc bag. Leave it for another 3 days.

6. Repeat steps 4 and 5 if your seeds have not germinated yet.

7. After germination has begun, plant the seeds in a planter of soil. Put it in a sunny spot. Water your plant every 3 days, and measure its growth.

8. Graph your observations of your plant's growth and make conclusions.

OTM2162 ISBN: 9781487710248
© On The Mark Press

Let's Observe

Illustration of the seeds after 3 days in the moist towel:

Illustration of the seeds after 6 days in the moist towel:

Illustration of the seeds after 9 days in the moist towel:

Tell about your seeds' germination process:

_____

_____

_____

_____

_____

_____

_____

OTM2162   ISBN: 9781487710248
© On The Mark Press

Graph the Growth!

Create a line graph to display the growth of your bean plant.

Let's Conclude

What conclusions can you make about your plant's growth?

_____

_____

_____

_____

OTM2162   ISBN: 9781487710248
© On The Mark Press

Let's Connect It!

Create a diagram to explain the life cycle stages of a bean plant.

At this stage...
_____
_____
_____
_____

The Seed

At this stage...
_____
_____
_____
_____

Flowering

Germination

At this stage...
_____
_____
_____
_____

The Plant

At this stage...
_____
_____
_____
_____

Challenge Question:

How could it continue to grow and spread in nature?

_____

_____

_____

OTM2162   ISBN: 9781487710248
© On The Mark Press

Germination Challenge!

Take the germination challenge! Choose your own seed type to germinate and grow into a plant. What are *you* going to grow?

Materials Needed:

- your choice of seeds
- a plastic Ziploc bag
- spray bottle of water
- a small gardening shovel

- water
- watering can
- a paper towel
- a planter with soil

What To Do:

1. Choose a seed type to germinate.

2. Using a spray bottle of water, mist a paper towel on both sides.

3. Put the seeds on the paper towel and fold it over a couple of times. Place it into the plastic Ziploc bag, and seal it.

4. Using worksheet 6, check and record the growth of your seedlings every few days. Be sure they are being kept moist!

5. Make comparisons of your bean plant's germination process to your new seeds' germination process. Record these on worksheet 6.

Name:

Observe & Compare

Illustration(s) of your seeds during the germination process:

Compare this seed type's germination process to the bean plant's germination:

_____

_____

_____

_____

_____

_____

_____

OTM2162   ISBN: 9781487710248
© On The Mark Press

Name:

Plant Adaptations

You have learned that plants need water, air, warmth, light, and space to grow. With these needs met, plants also create their own food through photosynthesis. But sometimes, even with these needs met, certain plants need to have special adaptations to help them survive and grow.

Sudan grass is poisonous only when wilted or frozen.

Plants, like animals, need to adapt to their environments. They do this in many different ways, whether it is against the colder seasons or to defend themselves against animals and people.

Deciduous trees lose their leaves in the winter. Many bushes and shrubs also lose their leaves when the frost comes. These plants go into a winter sleep, called **dormancy**. These plants may look dead in the winter, but they are not. It is just like people needing sleep to feel more energetic. Many plants that are dormant during winter need the coldness in order to grow again. For this reason, these plants will not grow in milder climates.

Plants cannot move to escape from predators, so some plants have special adaptations for survival, like poison! Their poisonous parts prevent them from being eaten by animals and insects. Poisonous plants have created a chemical inside their roots, stem, leaves, or

fruit that animals find distasteful or that will make the animal ill. Did you know that some of these plants are only poisonous at certain stages of their life cycles?

Poisonous plants are a danger to people too. These plants will make people ill if they are eaten or even touched. For example, poison ivy and poison oak cause an allergic reaction that produces an itchy rash.

OTM2162 ISBN: 9781487710248
© On The Mark Press

Research It!

Access the internet or visit your local library to learn about two plants with special adaptations. Name and draw each plant, then describe how each plant uses its adaptation to help it survive and grow in its environment.

Special adaptation:

_____

_____

_____

_____

_____

_____

_____

_____

Special adaptation:

_____

_____

_____

_____

_____

_____

_____

_____

OTM2162 ISBN: 9781487710248
© On The Mark Press

IMPORTANCE OF PLANTS

LEARNING INTENTION:

Students will learn about the importance of plants to people and other living things, and about the harmful effects plants face and how to minimize these effects.

SUCCESS CRITERIA:

- recognize that people get energy from food and that they release this energy
- use drawings and descriptions to communicate the way a food chain/ food web works
- make conclusions about the sun's energy as a food source for plants and make connections to animals and people in the environment
- design, plan, and construct a growing environment for plants
- make and record observations of the final product by comparing and sharing similarities, differences, and benefits of the design

MATERIALS NEEDED:

*Send a note home to parents explaining that their children are going to be building an item made from cleaned, recycled materials. A suggestive list of materials to have students bring in are: egg cartons, Styrofoam trays or aluminum plates, plastic bags, boxes, plastic bottles, milk jugs, margarine tubs or other plastic containers.

- scissors, glue, string, aluminum foil, duct tape, straws, skewers
- planting soil, small garden shovels, newspaper to cover desks
- assorted potting plants (about two for each student)
- chart paper, markers, pencils, pencil crayons, grid paper
- a copy of "Food Energy" worksheet 1 and 2 for each student

- a copy of "The Purpose of Plants" worksheet 3 for each student
- a copy of "What is the Harm?" worksheet 4 for each student
- a copy of "Lending a Hand" worksheet 5 for each student
- a copy of "A Growing Environment" worksheet 6, 7, and 8 for each student

PROCEDURE:

*This lesson can be done as one long lesson, or be done in five or six shorter lessons.

1. Engage the students in a physical activity to get their heart rates up. They can either take a short run, play tag, or skip for five minutes. After five minutes stop them and ask:
   - Where do you get the energy you need to exercise?
   - What do you put into your body to give you the energy to do this exercise?

   Explain to the students that our bodies burn food to get the energy we need to live and be active. All living things (plants, animals, and people) need food as a source of energy to grow and be active.

2. Using chart paper, create an example of a food chain, starting with a drawing of the sun. Ask students what uses the sun's energy to grow? (plants) Draw a plant. Ask students what insect/animal eats plants? Continue the questioning until the chain gets to a human at the top.

3. Break students into small groups, and give them a piece of chart paper and a marker. Instruct them to create a food chain. Once it is completed, have each group present their food chain to the larger group.

4. Broaden their knowledge by discussing food webs. Explain that all plants and animals are interconnected. Many connections exist within a small group of plants and animals. Using worksheet 1 and 2, students will create a food web. Upon completion, engage students in a discussion about plant and animal dependency by asking them, in what ways do animals help plants?

5. Give students worksheet 3. They will create a mind map to indicate ways that plants are useful to humans and other living things. A follow up option is to come together as a large group and record some of their responses on chart paper. This may help generate more ideas of their usage and importance, along with providing insight into the next activities.

6. Divide students into pairs and give them worksheet 4. They will engage in a 'think-pair-share' activity to discuss and record threats that plant life face.

7. Give students worksheet 5. Continuing to work with a partner, students will discuss things that they could do to minimize the harmful effects on plants so that they live and grow. Students can record their ideas inside the fingertips of the 'helping hand'.

8. Engage students in a discussion about things humans can do to ensure plants get their basic needs met and that they survive and grow healthy. Explain to students that they will design, plan, and construct an environment that helps plants to grow. Give them worksheets 6, 7, 8, and the recyclable and other assorted materials to construct. Upon completion of its construction, they will compare their final product with a partner and make some observations about its ability to meet the needs of plants.

DIFFERENTIATION:

Slower learners may benefit by:

• working with a partner to make the food web on worksheet 2

• pairing up with a strong peer to act as a recorder of ideas while completing worksheets 4 and 5

• working with a partner to plan and design a growing environment for plants, the construction could be completed as a team or independently

For enrichment, faster learners could:

• add plants and animals of their choosing to the food web on worksheet 2 by drawing them in the empty spaces on the page, and then draw their connections to the other plants and animals in the existing food web

• they could also choose to create another food web of their own

• using grid paper, they could design a floor plan of a greenhouse, calculate its perimeter and area

OTM2162 ISBN: 9781487710248
© On The Mark Press

Food Energy

You have learned that plants use energy from the sun to make food. They do this through photosynthesis. Plants depend on the sun for their survival.

Did you know that some animals get energy by eating plants? These animals depend on plants for their survival. Some animals get energy when they eat other animals. These animals depend on other animals for their survival. This is called a food chain. Food chains are all connected. A food web is a bunch of food chains. Let's create a food web!

Materials Needed:

- scissors • glue

What To Do:

1. Cut out the pictures on worksheet 1.

2. On worksheet 2, glue the pictures to make a food web.

3. Draw lines between plants or animals that eat each other.

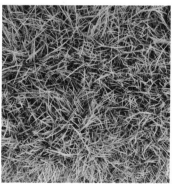

OTM2162   ISBN: 9781487710248
© On The Mark Press

Why is it called a food "web" and not a chain?

_____

_____

_____

_____

OTM2162　ISBN: 9781487710248
© On The Mark Press

Name:

The Purpose of Plants

Construct a mind map to show how plants are important to humans and other living things. Consider different points of view, for example, farmers, vegetarians, gardeners, scientists, wildlife, etc.

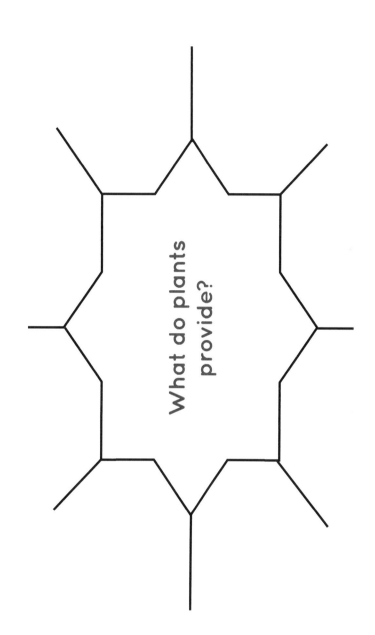

What do plants provide?

OTM2162   ISBN: 9781487710248
© On The Mark Press

Name:

What is the Harm?

Think **Pair** **Share**

With a partner, do some thinking and sharing of ideas about the threats to plant life. Consider such things as environmental conditions, animals, and human activities. Record your ideas in the box.

OTM2162   ISBN: 9781487710248

Lending a Hand

Discuss with your partner ways that *you* could reduce the harmful effects on plants and help them to survive.

Record your ideas inside the helping hands.

OTM2162   ISBN: 9781487710248
© On The Mark Press

Name:

A Growing Environment

Plants are important to us. Use what you know about plants to design and construct a healthy growing environment that meets all of their needs.

This is a design of my plant's growing environment:

This is a list of the materials I will use to build it:

- _____
- _____
- _____
- _____
- _____
- _____
- _____

OTM2162 ISBN: 9781487710248
© On The Mark Press

Name:

This is my plan for building the plant growing environment:

1. _____

_____

2. _____

_____

3. _____

_____

4. _____

_____

5. _____

_____

6. _____

_____

7. _____

_____

Now gather the materials to carry out your plan. Begin construction!

OTM2162 ISBN: 9781487710248
© On The Mark Press

Share and Compare

Working with a partner, compare your final product.

(A) Describe similar design techniques you used.

(B) Describe any differences in the design techniques you used.

Similarities	Differences

Let's Test It!

1. Plant some plants in your growing environment. Monitor them for a couple of weeks to ensure that their needs are being met.

2. Describe how your plants are doing:

_____

_____

_____

OTM2162 ISBN: 9781487710248
© On The Mark Press

ABORIGINAL PEOPLE AND PLANTS

LEARNING INTENTION:

Students will learn about the importance of some plants and their usage by First Nation's people.

SUCCESS CRITERIA:

- share prior knowledge of Aboriginal culture
- identify plants that are important to First Nation's people
- use drawings and descriptions to explain the importance or special usage of some plants in the Aboriginal culture
- make connections by explaining the importance of some plants to their own culture

MATERIALS NEEDED:

- a copy of "Grandma's Special Feeling" worksheet 1 for each student
- a copy of "The Four Sacred Plants" worksheet 2 for each student
- stories about plants and the Aboriginal culture *(see suggestions in #1 and #2 of procedure section)*
- access to the internet or local library
- white art paper
- pencils, pencil crayons, chart paper, markers

PROCEDURE:

*This lesson can be done as one long lesson, or can be done in two or three shorter lessons.

1. Engage students in a discussion about who First Nation's people are. Have students brainstorm some prior knowledge that they have about this culture, recording their ideas on chart paper. (A resource that could be shared with students is The Sharing Circle (Author: Theresa Meuse-Dallien).

2. Explain to students that plants are important in the Aboriginal culture. At this point, it may be beneficial to read aloud a story that explains the significance of plants to the Aboriginal people. A suggested book is Grandma's Special Feeling (Author: Karin Clark). After reading, ask students about some of the uses and importance of the plants to the Aboriginal culture. Give students worksheet 1 to complete.

3. Give students worksheet 2 to complete. Some students may need to access the internet for more information.

4. Engage students in a discussion about some plants that are important in their own culture, giving reasons/ explaining their use.

DIFFERENTIATION:

Slower learners may benefit by revisiting the story Grandma's Special Feeling, *while* they complete worksheet 1.

For enrichment, faster learners could make a cultural connection by choosing two plants that have an importance, or are of special use, to their own cultural background or of a culture of their choice. They can name them, illustrate them, and give reasons for their importance. A follow up option is to have these learners share their work with the large group.

Name:

Grandma's Special Feeling

In the story Grandma's Special Feeling, Grandma teaches her family about the many uses of plants and their importance to the First Nation's culture.

Name and draw two of the plants the children learn about on their outing. Explain the importance of these plants to the First Nation's culture.

Why is this plant important?

_____

_____

_____

_____

_____

_____

Why is this plant important?

_____

_____

_____

_____

_____

_____

OTM2162 ISBN: 9781487710248

The Four Sacred Plants

Use your own knowledge of the Aboriginal culture and knowledge from stories you have read to explain the importance of each of the sacred plants to First Nation's people.

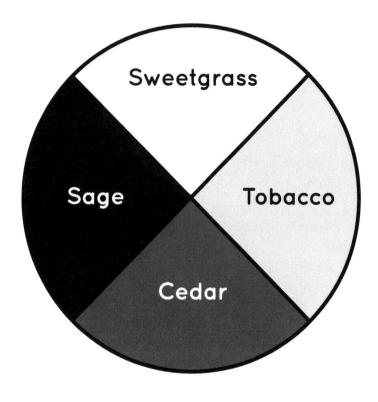

_____

_____

_____

_____

_____

_____

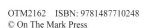

GROWING PLANTS FOR FOOD

LEARNING INTENTION:

Students will learn about the ways in which plants are grown for food, the advantages and disadvantages of organically grown and produced food, and harvesting methods of plants.

SUCCESS CRITERIA:

- identify places and ways in which plants are grown for food
- recognize the advantages of locally grown and organically produced food
- recognize the disadvantages of locally grown and organically produced food
- describe the growth process of an edible plant from seed to maturity using illustrations and written descriptions
- explain how and when an edible plant is harvested
- identify jobs that are related to the plant growing industry

MATERIALS NEEDED:

- a copy of "Growing Our Food" worksheet 1 for each student
- a copy of "The Good and the Bad" worksheet 2 for each student
- a copy of "From Seed to Harvest" worksheet 3 and 4 for each student
- access to the internet or local library
- pencils, pencil crayons, clipboards, chart paper, markers

PROCEDURE:

*This lesson can be done as one long lesson, or done in three or four shorter lessons.

1. Give students worksheet 1. They will create a mind map to indicate places/ ways in which plants are grown for food. Sample responses: on farms, in gardens, in greenhouses, in orchards.

2. Engage students in a discussion about the meaning of 'locally grown food' and organically produced food' to ensure their understanding of these concepts. Divide students into pairs and give them worksheet 2. They will engage in a 'think-pair-share' activity to discuss and record the advantages and disadvantages of growing food locally and organically produced food. Instruct them to record their thoughts in point form. A follow up option is to come back together as a large group to share, and record their responses on chart paper. This will promote some rich discussion. Pose this question: Have you considered the environmental benefits, if so what are they?

3. Give students worksheets 3 and 4. They will choose an edible plant, describe its growth process from seed to maturity, then describe its harvesting process. Some students may need to access the internet or local library.

4. As a large group, brainstorm some jobs that are related to the plant growing industry.

DIFFERENTIATION:

Slower learners may benefit by working as a small group with teacher support to complete worksheet 2. This would allow for probing of ideas and prompting of responses.

For enrichment, faster learners could research a job in the planting or farming industry. Questions to consider:

- What responsibilities does a person in this role have?
- What education or special knowledge is required to do this job?
- What type of equipment do they use? Describe how this equipment works.
- Would you like to do this job? Why or why not?

OTM2162   ISBN: 9781487710248
© On The Mark Press

Name:

Growing Our Food

Construct a mind map to show the different places or ways in which plants are grown for food.

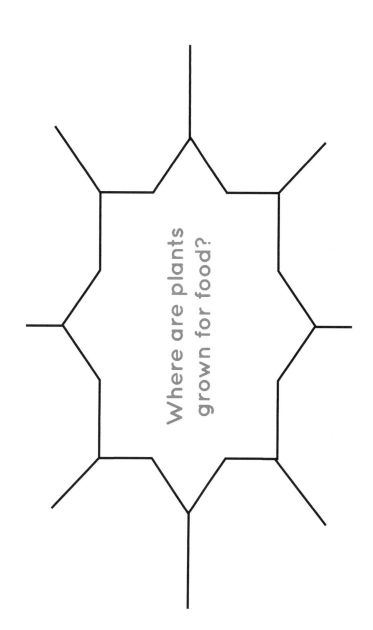

Where are plants grown for food?

OTM2162   ISBN: 9781487710248
© On The Mark Press

Name:

The Good and the Bad

The Advantages

With a partner, discuss the advantages and disadvantages of locally grown food and organically produced food. Record your thinking.

The Disadvantages

OTM2162 ISBN: 9781487710248

Name:

From Seed to Harvest

Choose an edible plant. Follow it from seed planting through its growing process, then to the final stage of harvesting.

Plant Name: _____

1. Does it grow from a seed or bulb?

   _____

2. What time of year is it planted?

   _____

3. Where can it be grown?

   _____

   _____

   _____

   _____

4. How long is its germination process?

   _____

   _____

5. How long does it take for it to mature and become ready for harvesting?

   _____

   _____

OTM2162   ISBN: 9781487710248
© On The Mark Press

Name:

Use the chart below to illustrate your edible plant's stages of growth.

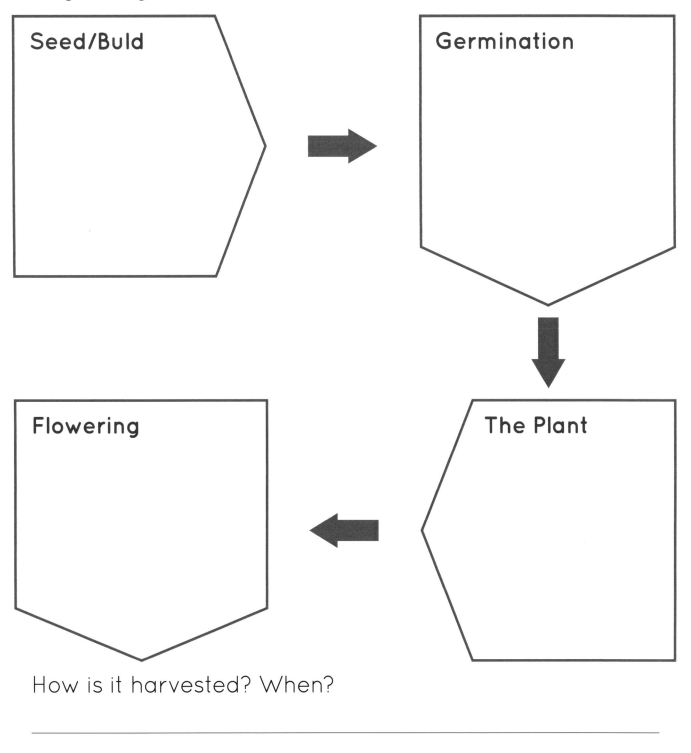

Seed/Buld

Germination

Flowering

The Plant

How is it harvested? When?

_____

_____

_____

_____

OTM2162 ISBN: 9781487710248
© On The Mark Press

INVERTEBRATES

LEARNING INTENTION:

Students will learn about the physical characteristics and behaviors of invertebrates.

SUCCESS CRITERIA:

- identify the body parts of an insect, a spider, a worm, and a crustacean
- choose an invertebrate to research
- describe its appearance, where it lives, what it eats, and special changes it experiences
- create a home for a hermit crab
- describe the needs of hermit crabs, and its growth and changes over time

MATERIALS NEEDED:

- a copy of "What's an Invertebrate?" worksheet 1 and 2 for each student
- a copy of "Investigating an Invertebrate!" worksheet 3 and 4
- a copy of "A Home for a Hermit Crab" worksheet 5 for each student
- access to the internet or local library
- chart paper, markers, pencil crayons, clipboards, pencils
- *visit a local pet store or go on line at www. petsmart.com to gather the materials and information needed in order to create a home for a hermit crab
- *painting paper, paint, paint brushes, modeling clay (optional materials)*

PROCEDURE:

*This lesson can be done as one long lesson, or be done in three or four shorter lessons.

1. Using worksheets 1 and 2, do a shared reading activity with the students. This will allow for reading practise and breaking down word parts to read the larger words. Along with the content, discussion of vocabulary words is beneficial to ensure comprehension.

Some interesting vocabulary words to focus on are:

- invertebrate
- glands
- crustaceans
- thorax
- burrows
- antennae
- spinnerets
- head
- cephalothorax
- exoskeleton
- arachnids
- vegetation
- abdomen
- moist

2. Students will brainstorm different invertebrates that they know. Record their responses. Instruct students to choose an invertebrate to research. Give them worksheets 3 and 4, and a clipboard and pencil. With access to the internet or local library, students will investigate and record information about their chosen invertebrate.

3. Together as a class, create a home for a hermit crab that can be maintained in the classroom. You will need to visit a local pet store, or go online at www.petsmart.com to gather the information on the materials you will need and how to care for it daily.

4. Once the hermit crab's home is created, give students worksheet 5. They will illustrate what it looks like. After a few weeks or months, they can provide detail on how to care for the hermit crab and comment on any changes they have noticed.

DIFFERENTIATION:

Slower learners may benefit by working in a small group with teacher support to orally answer the questions on worksheet 5. Responses could be recorded on one large chart paper, and then displayed in the classroom or on a bulletin board. Surround this chart paper with an art activity (painting) done by students that depicts the hermit crab in its environment.

For enrichment, faster learners could use modeling clay to create three dimensional models of the four types of invertebrates (insect, spider, worm, crustacean). These can be put on display.

OTM2162   ISBN: 9781487710248
© On The Mark Press

What is an Invertebrate?

An **invertebrate** is an animal with no backbone. Some invertebrates like **insects** have an **exoskeleton**, which is a skeleton that is on the outside of their bodies. Insects have bodies that are made up of three parts, the **head**, the **thorax**, and the **abdomen**. They have three pairs of legs that grow out from the front, middle, and back of the thorax. The head of an insect has two **antennae**.

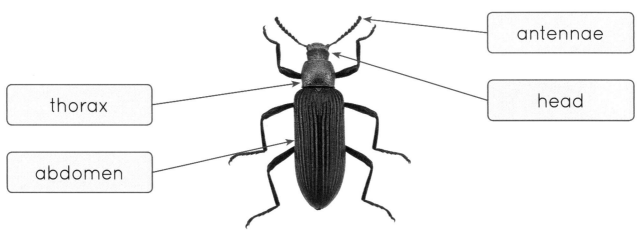

antennae

head

thorax

abdomen

Another group of invertebrates are **arachnids**. They have eight legs. They do not have wings or antennae. Can you guess what creature is part of this group? It is the spider!

Arachnids have bodies made up of two parts, the **cephalothorax** and the **abdomen**. The cephalothorax has the spider's eyes, brain, mouth fangs, and stomach. A spider's eight legs grow out of this part of its body. If a spider was poisonous, this is where its poisonous glands would be. A spider's **spinnerets** are at the back end of the abdomen. A spider makes **silk** in its spinnerets. Some spiders use this silk to spin webs.

As a spider grows, it sheds its first exoskeleton and then stretches out as a new, larger exoskeleton forms. Did you know that a spider's body has oil on it that helps it from sticking to its own web?

OTM2162 ISBN: 9781487710248
© On The Mark Press

Worms are also invertebrates. They have no legs, so they slither or inch their bodies along to move from place to place. There are different kinds of worms, but the one that is most common is the earthworm.

The earthworm's burrow is long and dark. Earthworms like to stay in their burrows because it is dark and moist.

Earthworms make their homes in soil. Some earthworms live in the roots of grass and some live in vegetable or flower gardens. Some earthworms will make their burrows under leaves around tree roots. They come out at night to eat leaves and grass, or other vegetation.

Another group of invertebrates are **crustaceans**. They live mostly in water. Some examples of crustaceans are lobsters, crabs, and shrimp.

lobster

crab

shrimp

Crustaceans have many legs with jointed parts. Their body is made up of the head, the thorax, and the abdomen. Their eyes are in the middle of their head. Did you know that crustaceans have two pairs of antennae? They use their antennae to sense what is in their environment, like food, or dangers.

OTM2162 ISBN: 9781487710248
© On The Mark Press

Investigating an Invertebrate!

You have learned about different types of invertebrates. Now it is time to choose one to learn more about what it looks like, where it lives, what it eats, and how it lives. Let's get started!

The invertebrate I am investigating is _____.

This is a diagram of what it looks like:

Where does it live?

_____

_____

_____

_____

OTM2162   ISBN: 9781487710248

Name:

Make a list of things that your invertebrate eats.

- _____
- _____
- _____
- _____
- _____

How does it get its food?

_____

_____

_____

_____

Tell about any special changes that happen to your invertebrate. For example:

• Does it migrate?

• Does it change its form?

• Does it molt?

_____

_____

_____

_____

OTM2162   ISBN: 9781487710248
© On The Mark Press

Name:

A Home for a Hermit Crab

Work together with your teacher and classmates to set up a home for a hermit crab. Watch for changes that happen as the hermit crab grows. Record what you see.

This is the home we made for the hermit crab:

How do you care for the hermit crab?

What growth and changes have you noticed in the hermit crab?

OTM2162   ISBN: 9781487710248
© On The Mark Press

BIRDS

LEARNING INTENTION:

Students will learn about the physical characteristics and behaviors of birds.

SUCCESS CRITERIA:

- identify the physical characteristics of birds
- choosing a bird to research, describe changes in its appearance as it grows
- describe where it lives, what it eats, and how it cares for its young
- display all information using pictures and written descriptions

MATERIALS NEEDED:

- a copy of "Birds" worksheet 1 for each student
- a copy of "Going on a Bird Study!" worksheet 2 and 3 for each student
- access to the internet or local library, pencils, pencil crayons
- sheets of art paper, paint, and paint brushes *(optional materials)*

PROCEDURE:

*This lesson can be done as one long lesson, or can be divided into two shorter lessons.

1. Using worksheets 1, do a shared reading activity with the students. This will allow for reading practise and breaking down of the larger words. Along with content, discussion of some vocabulary would be beneficial for comprehension.

Some interesting vocabulary words to focus on are:

- climates
- vertebrate
- hollow
- webbed feet
- skeleton
- clawed feet
- talons
- hatch
- migrate
- prey

2. Give students worksheets 2 and 3. With access to the internet or by visiting a local library, students will find out more about the physical appearance of the bird they chose and what it looked like upon hatching. They will also need to find out what it eats, and how it cares for its young. A follow up option is to have students showcase their picture of their bird to the large group and orally share one fact that they learned about it.

DIFFERENTIATION:

Slower learners may benefit by working as a small group with teacher support and direction to complete worksheets 2 and 3. Choosing one bird, assign a section to each student in the group. Then they can share their findings with the small group in order to complete the worksheets.

For enrichment, faster learners could paint a picture of their bird in its environment.

OTM2162 ISBN: 9781487710248
© On The Mark Press

Birds

Did you know that there are about 10 000 different kinds of birds in the world? They live all over the world, on almost every piece of land and in all climates. Some birds **migrate** from place to place. Other birds live in the same place year round.

Birds are **vertebrate** animals. This means they have a skeleton on the inside of their body. The inside of their bones are filled with tiny air sacs and their bones are hollow too. This makes them light enough to fly. The feathers on birds keep them warm and they also help them to fly. Birds get power and lift from beating their wings.

All birds lay eggs. A young bird grows inside the egg until it is ready to **hatch**. Young birds need a lot of care from their parents. The parent bird will catch insects and other food to feed it to their babies. When baby birds are older and have grown feathers and strong wings, the parents teach them to fly. Then they are ready to catch their own food.

While some birds eat things such as insects or worms, other birds eat small animals like mice, snakes, and even fish too! These birds are called **birds of prey**. They have **talons**, which are clawed feet made for grabbing prey.

Other birds like loons and ducks, live on water. They eat insects and fish. These birds have **webbed feet** to help them swim.

OTM2162 ISBN: 9781487710248
© On The Mark Press

Name:

Going on a Bird Study!

Choose a bird to study. Draw a picture of it that shows where it lives.

Draw what this bird looked like when it hatched as a baby from its egg.

How has it changed from when it hatched to now?

_____

_____

_____

OTM2162   ISBN: 9781487710248
© On The Mark Press

Name:

Draw pictures of some things that your bird likes to eat. Label them.

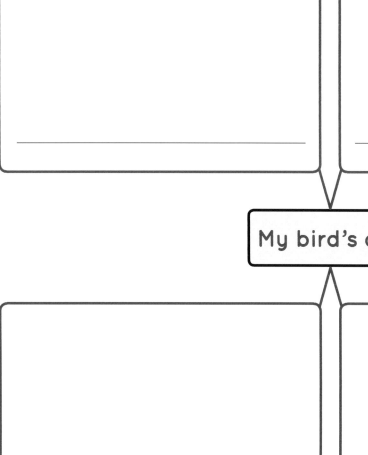

My bird's diet

Tell about some things that your bird does to care for its babies.

- _____

_____

- _____

_____

- _____

_____

OTM2162   ISBN: 9781487710248
© On The Mark Press

FISH, REPTILES, AMPHIBIANS

LEARNING INTENTION:

Students will learn about the physical characteristics of fish, reptiles, and amphibians; and research to learn more about their life cycles.

SUCCESS CRITERIA:

- recognize the physical characteristics of fish, reptiles, and amphibians
- identify reptiles and amphibians
- research an interesting fact about a chosen reptile or amphibian
- learn about and discuss the life cycle of a frog with a partner
- order the stages in a frog's life and describe each stage in words

MATERIALS NEEDED:

- a copy of "Fish" worksheet 1 for each student
- a copy of "Reading About Reptiles!" worksheet 2 for each student
- a copy of "About Those Amphibians!" worksheet 3 for each student
- a copy of "Life Cycle of a Frog" worksheets 4, 5, and 6 for each student
- additional copies of worksheets 5 and 6 (*optional materials*)
- access to the internet or local library
- pencils, pencil crayons, clipboards

PROCEDURE:

*This lesson can be done as one long lesson, or can be done in three or four shorter lessons.

1. Using worksheets 1, 2, and 3, do a shared reading activity with the students. This will allow for reading practise and breaking down of the larger words. Along with content, discussion of some vocabulary would be beneficial for comprehension.

Some interesting vocabulary words to focus on are:

- vertebrates
- oxygen
- hibernate
- scales
- blood vessels
- polar
- skeleton
- operculum
- nutrients

- tail
- carbon dioxide
- climate
- fins
- nocturnal
- reptile
- gills
- cold-blooded
- amphibian

2. Give students a list of reptiles and amphibians, (e.g., tortoise, gecko, newt, frog, cobra, lizard, iguana, salamander, turtle, toad, crocodile, caecilian). Allow them to access the internet to find out if they are either a reptile or an amphibian. As a follow-up option, they could choose one, find an interesting fact about it, and report it back to the large group.

3. Students can watch a video about the life cycle of a frog. A suggested video is 'Wild Kratts Aqua Frog', which can be accessed at www.youtube.com. Afterward, students can engage in a 'turn and talk' activity with a partner to discuss the information from the video that they found most interesting/ what they still wonder about. Give students worksheets 4, 5, and 6 to complete.

DIFFERENTIATION:

Slower learners may benefit by:

- working with a strong peer to research about reptiles and amphibians as per item #2
- working as a small group with teacher support to complete worksheets 4, 5, and 6. This would aid in the written description component, (an alternative is to eliminate the written expectation).

For enrichment, faster learners could chose a fish or reptile and investigate its life cycle. Giving them another copy of worksheet 5 and 6 would help to organize their learning.

OTM2162   ISBN: 9781487710248
© On The Mark Press

Name:

Fish

Did you know that there are about 28 000 different kinds of fish in the world? Fish can be found in lakes, rivers, oceans, streams, or ponds. Some live in fresh water and some live in salt water.

Fish are vertebrates. They have a skeleton inside their bodies. The outside of their bodies is covered in scales. They have fins and a tail which help them to swim in the water. Some fish have more fins than others.

Did You Know?

Most fish get their oxygen from water and not from air. Fish get oxygen from water through tiny blood vessels that are in their gills. The gills are under a protective flap called the **operculum**. There is a set of gills on each side of the fish's body.

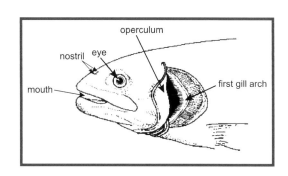

A fish opens its mouth to take in water. When its mouth is open, the operculum is closed. Then its mouth will close and the operculum will open.

The water that the fish took in its mouth passes over its gills and through its blood vessels. The fish uses the oxygen that is in the water to breathe. As water goes out the operculum, the fish breathes out carbon dioxide.

OTM2162   ISBN: 9781487710248
© On The Mark Press

Reading About Reptiles!

Did you know that there are over 8000 different kinds of reptiles in the world?

Tuataras prey upon insects, spiders, frogs, and other small reptiles.

Reptiles are animals such as snakes, lizards, turtles, crocodiles, alligators, and tuataras. Tuataras live in New Zealand. They are nocturnal reptiles. They sleep in the day and hunt for food at night.

Reptiles can be found all over the world, except in the polar areas because it is too cold there for them.

Reptiles are **vertebrate** animals. This means they have a skeleton on the inside of their body. Their skin is covered in scales, and some have a bony plate called a shell.

Reptiles are cold-blooded animals. This means that they cannot control their own body temperature. They need the heat from the sun to become warm and active. A reptile can get its warmth from the sun, heated rocks, logs, or soil.

FAST FACT!

Most reptiles lay eggs. When a baby hatches from its egg, it uses its egg tooth to break the shell open. Soon after, it loses its egg tooth.

OTM2162   ISBN: 9781487710248
© On The Mark Press

Name:

About Those Amphibians!

There are about 7000 different kinds of amphibians in the world. There are three groups of amphibians. The first group is tailless, like frogs and toads. The second group is tailed like salamanders and newts. The third group is called caecilians, these are worm-like creatures.

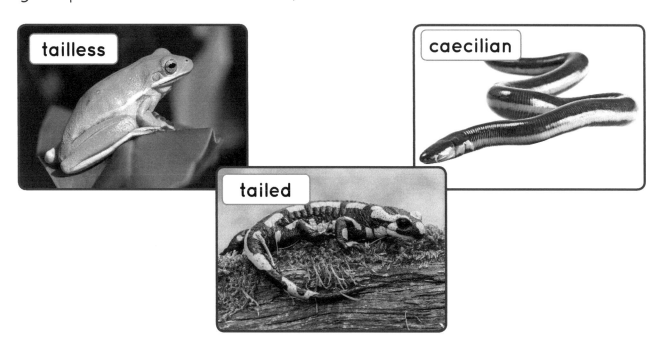

Amphibians are cold-blooded animals. They need the heat from the sun to warm up their body temperature to help them get active. Because of this, amphibians that live in cooler climates hibernate in the winter.

Amphibians are **vertebrate** animals. This means they have a skeleton on the inside of their body. Most amphibians have smooth, moist skin. There are a few, like the toad, that have scaly skin.

Amphibians often shed parts of their skin and grow new skin. Some will eat their skin for the nutrients in it. Did you know that some amphibians, like the salamander, can even grow new arms and legs if they lose one?

OTM2162   ISBN: 9781487710248
© On The Mark Press

Life Cycle of a Frog

Cut out the pictures below. On worksheets 5 and 6, order the pictures to show the life cycle of a frog. Beside each picture, write a sentence to explain what is happening at each stage of its life cycle.

Name:

_____

_____

_____

_____

_____

_____

_____

_____

_____

_____

_____

_____

_____

_____

_____

OTM2162   ISBN: 9781487710248

OTM2162   ISBN: 9781487710248
© On The Mark Press

MAMMALS

LEARNING INTENTION:

Students will learn about the different categories of mammals and their physical characteristics; then research to learn more about a mammal of their choice.

SUCCESS CRITERIA:

- identify three categories of mammals, describe their physical characteristics
- list different types of mammals
- choosing a mammal to research, describe how it is born and nurtured by its mother
- describe what it eats, how it moves, where it lives
- identify its predators
- describe any special changes the mammal experiences
- display all information using pictures and written descriptions

MATERIALS NEEDED:

- a copy of "Mammals" worksheet 1 for each student
- a copy of "A Mammal Study" worksheets 2, 3, and 4 for each student
- access to the internet or local library
- pencils, pencil crayons, clipboards

PROCEDURE:

*This lesson can be done as one long lesson, or done in two or three shorter lessons.

1. Using worksheets 1, do a shared reading activity with the students. This will allow for reading practise and breaking down of the larger words. Along with content, discussion of some vocabulary would be beneficial for comprehension.

Some interesting vocabulary words to focus on are:

- egg-laying
- pouch
- placental
- marsupial
- control
- nourish
- protection
- appendages
- blindly
- warm-blooded
- vertebrate
- underdeveloped

2. Engage students in a brainstorming discussion about the common characteristics of mammals. Record students' responses in a mind web on chart paper. A completed mind web should include these ideas:

- have fur or hair
- feed their young milk
- give their young care and protection
- are warm-blooded
- have a backbone
- have a skeleton inside their body
- have 4 appendages
- have lungs and need air to breathe

3. Continue to brainstorm as a large group, ask students to name some mammals. Record their responses on chart paper.

4. Give students worksheets 2, 3 and 4. With access to the internet or by visiting a local library, students will find out more about a mammal of their choice.

DIFFERENTIATION:

Slower learners may benefit by working as a small group with teacher support to complete worksheets 2, 3, and 4. Choosing one mammal, assign a section to each student in the group. Then they can share their findings with the small group in order to complete the worksheets.

For enrichment, faster learners could find out how long their chosen mammal lives in captivity (e.g., in a zoo), and in the wild. They could provide an explanation as to why there would be a difference in the life expectancy.

OTM2162   ISBN: 9781487710248

Name:

Mammals

Did you know that there are about 5400 different kinds of mammals in the world? These mammals can be divided into three main groups. There are egg-laying mammals, marsupials, and placental mammals.

Placental mammals give birth to live young. Mammals are the only animals that can feed their young milk. When a mammal is born, it is not able to take care of itself. It needs its mother for food and protection.

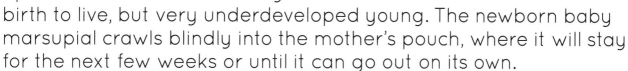

Egg-laying mammals, such as the platypus and the echidna, lay eggs instead of giving birth to live young. When platypus babies hatch out of their eggs their mothers nourish their young with milk. They feed them for about three months or until they swim on their own.

Marsupials are animals such as koalas, kangaroos, and opossums. These mammals give birth to live, but very underdeveloped young. The newborn baby marsupial crawls blindly into the mother's pouch, where it will stay for the next few weeks or until it can go out on its own.

Mammals are **warm-blooded** animals. This means that they can control their own body temperature. They are **vertebrate** animals so they have a backbone and a skeleton on the inside of their body. They have lungs and need air to breathe. Mammals usually have four appendages (arms and legs). All mammals are covered in fur or hair which keeps them warm.

DID YOU KNOW?

Humans are mammals! We are placental mammals. Humans give birth to live young. We need our mothers for food and protection when we are young, until we learn to care for ourselves.

OTM2162   ISBN: 9781487710248
© On The Mark Press

A Mammal Study

Visit the library or access the internet to find out some interesting facts about a mammal that you would like to learn more about.

Choose It!

The mammal that I would like to learn more about is the

_____.

Describe It!

Illustrate the mammal that you chose.

What type of mammal is it? (circle one)

egg-laying   **marsupial**   **placental**

OTM2162   ISBN: 9781487710248
© On The Mark Press

Name:

Study It!

How long does it take for it to hatch or be born?

_____

_____

_____

_____

How long does it stay with its mother?

_____

_____

_____

_____

What does it eat?

(a) when it is a baby - _____

(b) when it is a young mammal - _____

(c) when it is an adult - _____

Circle all the ways that your mammal moves.

walk	run	hop
swim	fly	crawl

OTM2162   ISBN: 9781487710248
© On The Mark Press

Name:

(A) Draw your mammal where it lives.

(B) Draw and label some of its predators that live there too.

Tell about any special changes that happen to your mammal. For example:

• Does it migrate?

• Does it hibernate?

• Does it camouflage?

_____

_____

_____

OTM2162 ISBN: 9781487710248
© On The Mark Press